Sustaining Air Force Space Systems

A Model for the Global Positioning System

T0195517

Don Snyder · Patrick Mills · Katherine Comanor · Charles Robert Roll, Jr.

Prepared for the United States Air Force

Approved for public release, distribution unlimited

RAND PROJECT AIR FORCE

The research described in this report was sponsored by the United States Air Force under Contract F49642-01-C-0003 and FA7014-06-C-0001. Further information may be obtained from the Strategic Planning Division, Directorate of Plans, Hq USAF.

Library of Congress Cataloging-in-Publication Data

Sustaining Air Force space systems : a model for the Global Positioning System / Don Snyder ... [et al.].
    p. cm.
    Includes bibliographical references.
    ISBN 978-0-8330-4044-2 (pbk. : alk. paper)
    1. Astronautics, Military—United States—Equipment and supplies. 2. Global Positioning System. 3. United States. Air Force—Procurement. 4. United States. Air Force Space Command—Planning. I. Snyder, Don, 1962– II. Rand Corporation.

UG1523.S87 2007
358'.88—dc22

2007001853

The RAND Corporation is a nonprofit research organization providing objective analysis and effective solutions that address the challenges facing the public and private sectors around the world. RAND's publications do not necessarily reflect the opinions of its research clients and sponsors.

RAND® is a registered trademark.

Published 2007 by the RAND Corporation
1776 Main Street, P.O. Box 2138, Santa Monica, CA 90407-2138
1200 South Hayes Street, Arlington, VA 22202-5050
4570 Fifth Avenue, Suite 600, Pittsburgh, PA 15213-2665
RAND URL: http://www.rand.org/
To order RAND documents or to obtain additional information, contact
Distribution Services: Telephone: (310) 451-7002;
Fax: (310) 451-6915; Email: order@rand.org

Preface

Air Force Space Command (AFSPC) needs quantitative tools to assist it in making decisions on how changes in the dollars invested in maintenance and sustainment of the ground segment of space systems affect the operational performance of those systems. This monograph outlines criteria for analyzing how sustainment investments affect the operational performance of space systems, focusing on the Global Positioning System. We offer a framework for such analyses and recommend steps to implement that framework.

The research reported here was sponsored by Air Force Space Command. The work was conducted within the Resource Management Program of RAND Project AIR FORCE as part of a project begun in late fiscal year 2005, "Air Force Space Command Logistics Review." A related document is

- *Space Command Sustainment Review: Improving the Balance Between Current and Future Capabilities*, Robert S. Tripp, Kristin F. Lynch, Shawn Harrison, John G. Drew, and Charles Robert Roll, Jr. (MG-518-AF, forthcoming).

The research for this report was completed in February 2006.

RAND Project AIR FORCE

RAND Project AIR FORCE (PAF), a division of the RAND Corporation, is the U.S. Air Force's federally funded research and development center for studies and analyses. PAF provides the Air Force with independent analyses of policy alternatives affecting the development, employment, combat readiness, and support of current and future aerospace forces. Research is conducted in four programs: Aerospace Force Development; Manpower, Personnel, and Training; Resource Management; and Strategy and Doctrine.

Additional information about PAF is available on our Web site at http://www.rand.org/paf.

Contents

Figures

Summary

Aging systems and systems operating longer than their anticipated life span, sometimes because of program slips in follow-on systems, have intensified the need for understanding how maintenance and sustainment affect the performance of space systems. In this monograph, we develop a pilot framework for analyzing these and related questions in the ground segment of the Global Positioning System and recommend steps for implementing this framework. In doing so, we address the issue of modeling approach and how to define appropriate metrics of performance. We develop the guidelines for metrics and analytic methods as generally as possible so that they will be useful for other space systems.

Much of the spirit of the current metrics used to monitor the maintenance of the ground segments of space systems follows that of metrics used for aircraft. But, space systems have some attributes that differ significantly from those of aircraft systems, and these attributes suggest that the metrics for maintenance and sustainment for space systems be reconsidered. From a modeling perspective, the central difference is that space systems are highly integrated systems in near constant operation, not fleets of aircraft, any one of which can perform the specified mission. This difference leads to three challenges for the analyst.

First, the logical metric used in the aircraft realm—the fraction of the fleet that can perform the stated mission—is not applicable in the space realm. Space command systems function as an integrated whole, and the whole must meet operational mission goals at all times.

What is needed for space systems is either a measure or measures that reflect the overall system performance, even when the system is operating nominally. The metric should also be sensitive to sustainment perturbations. We call a measure of performance that has these qualities a *sentinel metric*. A further constraint on performance-metric selection is that the users of space systems are often diverse, spanning the various military services, other governmental organizations, and, even occasionally, the civilian sector. Each of these users may require different capabilities and levels of performance to satisfy their own mission requirements.

Second, for the ground segments of most space systems, what makes components break—and a related matter, what modifications make components more reliable—are not as well understood as cause-and-effect linkages are in the aircraft domain. Flying hours drive some engine maintenance in jets, but what preventative maintenance efforts lead a software-dominated system to be more reliable? When does maintenance intervention in software introduce bugs that lower system reliability in the short term, and when should such intervention be avoided?

Third, even when causal linkages are understood, since space systems are operated as single entities and not as sets of individual capabilities, there are many fewer identical components and failures from which to collect statistically meaningful data. If the statistical distributions of underlying data, such as the time between failures and the time to restore function, are not well constrained, the fidelity of the predictive estimates of performance diminishes.

For a pilot study, we examine the Global Positioning System (GPS) and how a model might be developed to explore how programming[1] investments and trade-offs in maintenance and sustainment for the ground segment of this system might be analyzed. The GPS is a satellite-based system that provides accurate spatial location and timing data for civilian and military users. It is composed of three segments:

---

[1] Unless otherwise indicated, we use the terms *programming* and *programmer* to refer to the activities and individuals involved in the building of the Air Force Program Objective Memorandum (POM), not to computer code.

the user segment, receivers that GPS users employ to locate themselves and determine time; the space segment, the satellite constellation; and the ground control segment, which will be the focus of this study. The ground segment has three subsystems: monitoring stations, the Master Control Station, and ground antennas. One of the main functions of the ground segment is to monitor and maintain the accuracy of the overall system. The monitoring stations check on the status of the satellites, the Master Control Station makes decisions on updates to the satellites, and the ground antennas transmit those updates to the satellites (see pp. 7–12).

The starting point for modeling the effect of sustainment activities on operational performance is the selection of a measure of performance. The qualities of the measure of performance determine the scope of the decisions that can be made using the model, and they dictate the minimum level of granularity of the data-collection and analysis efforts. For the GPS program, regardless of the user, the appropriate sentinel measures of performance are measures of the variance over time of the accuracy of the user's location and time estimates. These broad metrics are appropriate for programming decisions, and they may differ from metrics used to determine operational priorities.

We examine the effect of the reliability of one subsystem of the GPS ground segment, the ground antennas, on the variance over time of the accuracy of a user's location estimate. Specifically, we examine a proxy for this measure: What is the approximate difference in where the satellites are relative to where they appear to be to a user (called the *estimated range deviation* [ERD]), averaged over the satellite constellation. Three types of service disruptions of ground antennas affect this measure: unscheduled maintenance, scheduled maintenance, and interruptions in the communications links connecting the ground antennas to the Master Control Station.

Scheduled maintenance includes all maintenance activities that are done on a regular basis, along with installation of system-component upgrades. *Unscheduled maintenance* includes hardware breaks, electronic component failures, and software crashes. Failures in communications links between the subsystem and the Master Control Station fall under the purview of, and are maintained by, the Defense Information Systems Agency (DISA), which is outside the control of Air Force Space Command. Nevertheless, these outages need to be quantitatively understood and included as part of the model so that the limits of Air Force actions on the system performance are understood.

Each subsystem is composed of a multitude of parts, and each part will have times between breaks that can be described by some probability density function. Once broken, each component requires some time before its function is restored that is also described by some probability density function. This time is the sum of the time to repair the component and any time that it takes to get that component and the maintenance personnel to the site.

The system can be modeled by collecting and analyzing the failure rates and restoration times of each of the components. However, such an analysis alone will not capture the full behavior of the system. Evaluating the performance of a system requires a systemwide view that incorporates not only the performance of the components but how they mutually interact, how they communicate with one another, redundancies, and the overall command and control of the system. For this reason, evaluating how maintenance and sustainment efforts affect space system performance should start with a systemwide view and work down to individual maintenance and sustainment activities (a top-down approach) (see pp. 12–19).

Using a top-down approach does not invalidate the need for an understanding of component-level failures. Rather, the systemwide, operational view places the components in context and reveals a priority for data collection and analysis. That is, a systemwide view indicates which subsystems or components are most problematic and, hence, are deserving of the highest level of attention in failure and repair data-collection and analysis. Once the key problems are identified, whether they are components failing, communications-link failures, lack of redundancy, or other issues, data for costs to remediate the problems can be estimated by examining their service-interruption modes in detail. This detail ties dollars invested to overall system performance as measured by the user's needs.

A complete, predictive analysis of maintenance and sustainment efforts for space systems then unfolds in the following steps. First, the operational objectives of the users are quantified in a way that reflects the long-term behavior of the system that is likely to be affected by programmatic decisions. These *operational* objectives then define the metrics for *maintenance and sustainment*. A predictive model based on a systemwide view links the maintenance and sustainment efforts to the operational metrics. This predictive model then reveals critical problem areas, which can be explored in greater detail. Once the critical areas are identified, additional analysis at the component level then links the remedies with costs, indicating how investments in resources affect operational performance.

For these reasons, in this monograph we start with a top-down approach to modeling the GPS (see pp. 12–19, 21–24). This approach puts the perspective of the user in the forefront, thereby placing the user's priorities in a position to motivate the maintenance and sustainment metrics. Although the scope of this study limited us from linking this work to component-level analysis and, hence, directly to costs, the approach explored in this monograph complements ongoing component-level analysis being done by Air Force Space Command (AFSPC/A4S). Linking the analysis presented here with ongoing work at AFSPC can present a complete, predictive model of space systems that reveals how dollars allocated in the budget affect the overall space system in terms of operational (not maintenance) performance.

Preliminary results indicate that, when ground antennas' reliability is considered in isolation, significant operational-performance deterioration will occur when the mean time between failures of ground antennas is less than 15 hours (given 5 hours for mean time to restore function) and when the mean time to restore function exceeds about 20 hours (given 50 hours for mean time between failures). Adding an antenna adds redundancy to a redundant system, providing little additional accuracy unless maintenance is quite poor. If system performance is to remain nominal, losing an antenna requires exemplary maintenance on the remaining antennas. (See pp. 24–35.)

The logical steps for implementing the framework developed in this report are as follows:

- Expand the model to include the reliability of the monitoring stations and that of the Master Control Station (and its backup facility). (See pp. 37–38.)
- Collect comprehensive data on when each of the subsystems is not functioning well enough to perform its assigned mission. This collection effort should include instances when the software crashes and needs to be reset, as well as such factors as failures of the communications links, even if these factors lie outside the control of AFSPC, and any other times (of which we are unaware) that a subsystem is operationally unavailable. This data-collection effort should be prioritized by system-level analysis of how maintenance affects the various users' requirements. (See pp. 38–39.)
- Extend the study to targeted components, to include the relationship of dollars invested into sustainment to the probability distributions of break rates and time to restore function. Key issues are, What causes breakages of mechanical components? Failures of electrical components? and Changes in software reliability? Specifically, are system failures correlated with service cycles, duration of use, or other factors? And what are the consequences of deferring scheduled maintenance on these systems to future break rates and break types? (See p. 39.)
- Expand the analysis to include other ways of increasing system performance, including improving the quality of the GPS algorithms, introducing more-advanced technologies, and providing cross-link capability among the satellites. (See pp. 39–40.)
- Expand the analysis to examine how fast the performance of the system degrades in response to an abrupt decrease in maintenance performance (i.e., the relaxation times of the GPS to perturbations in mean time between critical failures and mean time to restore function). (See pp. 31–32, 40.)
- Expand the analysis to embrace other space systems. (See pp. 40–41.)

Acknowledgments

This work would not have been possible without the support of many individuals. At Air Force Space Command Headquarters, we especially thank Col Samuel Fancher, Brian Healy, Chris Milius, and MSgt Thomas Oaks; at the Space and Missile Systems Center, we thank Louis Johnson, Trenton Darling, and Tim McIntire; at the 2nd Space Operations Squadron, we thank Maj Theresa Malasavage, Philip J. Mendicki, and Brian Brottlund; at the 19th Space Operations Squadron, we thank Maj James Pace; and at the Global Positioning System program office, we thank Col Rick Reaser, Col Kenneth Robinson, and Robin Pozniakoff.

At RAND, we are indebted to a number of researchers for a myriad of discussions and for reading drafts of various parts of this monograph. They are (in alphabetical order): Mahyar Amouzegar, Lionel Galway, David George, Lt Col Shawn Harrison, Richard Hillestad, Lance Menthe, Louis Miller, Adam Resnick, Lara Schmidt, and Robert Tripp. Reviews by Bernie Fox and Mel Eisman improved the monograph substantially. The authors remain responsible for all errors and omissions.

Abbreviations

AFSPC	Air Force Space Command
DISA	Defense Information Systems Agency
ERD	estimated range deviation
GPS	Global Positioning System
JDAM	Joint Direct Attack Munition
MCS	Master Control Station
MTBCF	mean time between critical failures
MTTRF	mean time to restore function
NATO	North Atlantic Treaty Organization
NGA	National Geospatial-Intelligence Agency
POM	Program Objective Memorandum
URE	user range error

CHAPTER ONE

Introduction

In times of constrained budgets and competing priorities, planners and programmers must understand how much the capability of a system will change in response to variations in the budget appropriated to an element of that system. Specifically, the following questions arise: How much additional capability is realized by increasing the budget by a certain amount? and, conversely, How much risk is assumed by decreasing the budget? In many areas of procurement, techniques in cost analysis shed considerable light on these relationships. But many other budgeting decisions pose considerable challenges. One such decision is how variations in maintenance and sustainment investments affect operational performance in a program, both in the short term and over longer terms. How changes in sustainment investments affect operational performance in aircraft systems can be difficult to quantify, but such analyses are yet more challenging in the ground segments of space systems under the purview of Air Force Space Command (AFSPC).

Within AFSPC, the approach adopted for measuring and reporting the performance metrics of efforts to sustain and maintain the ground segments of space systems is similar to those used to monitor Air Force aircraft. Some of these metrics include how frequently parts break, how fast those broken parts can be repaired, and what fraction of time the overall system is functioning nominally. That such measures are used is not surprising. Many of the maintenance officers in AFSPC spend substantial time in the aircraft side of the Air Force, and they are accustomed to this perspective. Further, these metrics capture some obviously important characteristics of any system. But, space systems possess some attributes that differ significantly from those of aircraft systems, and these attributes suggest that the metrics for maintenance and sustainment for space systems should be reconsidered.

In aircraft systems, the link between servicing and sustainment activities and operational performance measures has been reasonably well established. The operational goal is fairly well captured by the measure of what fraction of the fleet is capable of performing its assigned mission at a given time. The sustainment efforts largely consist of the inspecting, troubleshooting, removing, replacing, and repairing of parts. Years of experience have revealed how aircraft activities drive sustainment efforts. For example, some parts (e.g., jet engines) are known to require scheduled maintenance in proportion to flying hours, others (e.g., brakes and tires) in proportion to takeoffs and landings.

Identifying and quantifying these cause-and-effect linkages indicate what data need to be collected. With these data and linkages, analysts can estimate the sustainment demands (costs) given certain operational tempos. Models have been built that exploit this knowledge to anticipate future sustainment costs. Further, constrained part supplies affect aircraft mission-capable rates directly. This relationship provides an opportunity to model how changing maintenance and sustainment practices might impact the ability to generate aircraft sorties.

These characteristics of aircraft differ significantly from those of most of the ground systems maintained by AFSPC that monitor and communicate uploads to satellites. From a modeling perspective, the central difference is that the overall space systems are not sets of resources (fleets), each element of which performs a specified mission, leading to a logical measure of performance of what fraction of that set (fleet) can perform the stated mission. Space command systems generally function as an integrated whole, and the whole must meet operational mission goals at all times. Although the analogy is imperfect, space systems resemble a single aircraft more than they do a fleet of aircraft.

An aircraft can either perform or not perform an assigned mission, depending on the health of all its components. Redundancy in some aircraft components may prevent failure of a single component from jeopardizing the entire system. Likewise, space systems can often continue to satisfy mission capabilities through component failures, thanks to redundancy. But, unlike an individual aircraft, space systems must function continuously, even during times of maintenance.

Challenges to Space System Modeling Efforts

These key differences between aircraft and space systems present significant challenges that have impeded efforts to model the effect of sustainment investments on system performance. We highlight three of these challenges:

First, metrics for expressing operational capability for space systems are not as evident as those for aircraft. Merely measuring whether a space system performs its assigned mission is not a sufficiently demanding measure. Many space systems have, according to national strategic priorities, always performed their assigned mission. They have done so despite variations in the health and status of their subsystems, thanks to the redundancy of those subsystems. Measuring operational performance by whether or not an overall system performs its mission objectives is, therefore, an anemic predictor of future system performance. A system's latent ability to perform its mission might deteriorate over time, yet subsystem redundancy might shield this atrophy from affecting a performance metric based on whether the overall system is functioning according to its assigned mission. Hence, when the deterioration reaches the point at which the redundancy fails, the system will fail catastrophically and the performance measure will have failed to give sufficient forewarning to programmers to act to stave off the catastrophic failure.

What is needed is a measure (or measures) that alerts decisionmakers of trouble in time to anticipate problems programmatically. Such a metric must be sensitive to sustainment perturbations, such as the effect of variations in parts supplies, so that modeling with this metric can reasonably predict the implications of sustainment efforts on the future health of the overall system. In particular, this metric must vary measurably in response to variations in factors of interest in such a way that permits the identification of the point at which the system will fail catastrophically if the status quo continues. We call a metric with these qualities a *sentinel metric.*

A further constraint on performance-metric selection is that the users of space systems are often diverse, spanning the various military services, other governmental organizations, and even, occasionally, the civilian sector. So, even when measures have been defined that are appropriately sensitive to variations in sustainment efforts, these measures must also capture the various needs of this disparate group of users.

Second, for most space systems, which factors determine when components break—and a related matter, which modifications make components more reliable—are not as well understood as cause-and-effect linkages are in the aircraft domain. Flying hours drive some engine maintenance in jets, but what preventative-maintenance efforts cause a software-dominated system to be more (or less) reliable? When does maintenance intervention in software introduce bugs that lower system reliability in the short term, and when should such intervention be avoided?

Third, even when causal linkages are understood, space systems are single entities and not sets of individual capabilities. Therefore, there are many fewer identical components and failures on which to collect statistically meaningful data. A model is no more reliable than the data it processes. If the statistical distributions of key underlying data, such as time between failures and time to restore broken parts to their nominal function, are not well constrained, the fidelity of the model results diminishes.

Yet space systems are central to the warfighting effort, and failure of some of the systems could be catastrophic to that effort. As space systems age, and because some are performing for longer than anticipated—sometimes as a result of program slips in follow-on systems—such questions as what levels of sustainment are necessary to avoid such failures have become acute. The nation needs more than trailing indicators of space system performance; it needs modeling strategies based on sentinel metrics for predicting operational performance, given variations in sustainment efforts.

In this monograph, we develop a pilot framework for analyzing these and related questions for the ground segments of space systems. In doing so, we stress the need to adopt a systemwide view of the readiness of space systems and to link the effect of sustainment and maintenance efforts of the ground segments to the overall operational function of the system. Intimately associated with this framework is how to define appropriate measures of performance for these purposes. We discuss the attributes of such sentinel metrics and how they differ from metrics conceived to support operational decisions.

For a pilot study, we examine the Global Positioning System (GPS) and how a model might be developed to explore analysis of programming investments and trade-offs in sustainment and maintenance in this program. Within this system, we focus on one subsystem: the set of ground antennas used to broadcast signals to the satellite constellation. We examine just this one subsystem for simplicity and because the scope of this study is limited. Despite the focus on a subsystem, a system view is maintained throughout. Using this subsystem, we illustrate the modeling approach, and then indicate how the whole system might be analyzed similarly. By looking at a specific space system (in this case, GPS), we can discover problems and obstacles to analysis that abstract reasoning alone might miss, and we can reveal specific steps toward implementing a programming decision-support tool.

Why the GPS?

For many reasons, we chose the GPS program in particular for a case study. This system possesses many of the complexities of space systems in general that were mentioned above; hence, it is fertile for exploring the various difficulties of modeling these issues in space systems. Specifically, (1) the GPS has numerous users spanning the military and civilian sectors; (2) failure of the GPS to function *continuously* to specifications would have severe implications for national interests; (3) maintenance data on GPS components are sparse; and (4) the mathematical relationships between sustainment efforts and operational performance remain unclear.

Despite these difficulties, the GPS program has some characteristics that facilitate modeling, most important among them being that it delivers products that can be well defined quantitatively: time and the geographic position of a user. These well-defined characteristics simplify the problem of defining a useful, sharp measure of operational performance. Also, varying sustainment effort will cause these characteristics to vary measurably, making them sentinel metrics. Beyond this attribute, the GPS program is fairly self-contained, and relative to many other space systems, its various parts provide clear, distinct roles in maintaining the overall system's ability to provide accurate location and timing information.

The scope and time constraints of this study have not allowed us to explore programming trade-offs to firm conclusions even within the GPS program. We would be remiss, nevertheless, not to reflect on the degree to which common approaches to these three problems might work across AFSPC systems. This point is important, because not recognizing the uniqueness of each system can lead to analyses that fail to capture the essential elements of each system. Yet failure to define common measures and standards of analysis across systems can lead to confusion, fail to leverage economies of effort,[1] and hinder the ability to evaluate programming trade-offs.

Organization of This Monograph

The remaining chapters of this monograph describe the Global Positioning System at a level of detail needed for the analysis here (Chapter Two); discuss how to approach modeling the relationships between sustainment activities and overall system performance, and describe a pilot model for such analysis (Chapter Three); and examine the results of this model and how they might be used in policy analysis and, finally, discuss the implications for developing such models in GPS and other programs (Chapter Four).

---

[1]   For example, having similar standards for reporting criteria across systems can facilitate automated data collection.

CHAPTER TWO

Considerations for a GPS Sustainment Model

This chapter discusses the range of considerations for modeling the ground support of space systems, with a particular view toward the GPS. It begins with an overview of the GPS, paying particular attention to the role of the ground segment. That description provides the necessary background for a discussion of the attributes that a sustainment model of the GPS should possess. The chapter concludes with the description of a prototype of such a model that links maintenance performance with operational performance.

An Overview of GPS

The Global Positioning System is a satellite-based space system that provides accurate location and timing data for civilian, military, and nonmilitary governmental users. Although the Department of Defense provides this service, users are responsible for purchasing their own receivers.

From an Air Force perspective, the GPS is composed of three parts, generally called segments[1]: the constellation of satellites, or the space segment; a ground control segment; and a user segment. The *user segment* consists of the set of military GPS receivers that provide time and location for the services. Although this segment is a substantial part of the GPS system and budget, responsibility for it falls largely

---

[1] The GPS also provides a nuclear-detection capability that will not be discussed in this monograph.

under the purview of the GPS Joint Program Office and is beyond the scope of this study. Here, we analyze how the accuracy of the overall system is sustained, an activity that primarily involves the ground segment and the satellite constellation. In this section, we first describe the satellite constellation and the basics of how the GPS works, followed by an overview of the ground segment. The discussion focuses on those aspects of these segments that play a role in how sustainment affects overall system performance.

The space segment consists nominally of 24 satellites[2] distributed in six different orbital planes inclined to the Earth's equator by 55 degrees (deg). Each satellite completes an orbit in approximately 12 hours (hr). This configuration ensures that at least four satellites are in view at all times from any location on Earth, thus allowing a user at any terrestrial location to determine the time and the three spatial coordinates of position. A GPS receiver calculates the local position by determining the phase shift needed to match a pseudo-random code in the receiver with an identical one broadcast by a satellite. This phase shift gives the transit time for the signal, and, from the speed of the electromagnetic wave, the distance to the satellite (called the pseudo-range). From knowledge of the position (ephemeris) of the satellites, the position of the user is fixed by triangulation on multiple satellites.

Given the high speed of electromagnetic radiation, accurate timing is critical to the position calculation. Each GPS satellite has atomic clocks onboard for timing, but to keep the costs of GPS receivers reasonable, receivers contain less-accurate clocks than the satellites. The user equipment solves for time in addition to the three spatial coordinates, and these four unknowns require at least four satellites to be in view. The greater the angular spread of these satellites relative to the user, the more accurate the triangulation. Any additional satellites within view add degrees of overdetermination, thereby improving the accuracy of the user's time and position estimates.

The satellites continuously transmit information on two carrier signals, designated L1 and L2. L1 is modulated with a short-cycle-

---

[2] The 24 satellites are supplemented with backup satellites for redundancy. Currently, 29 satellites are in orbit.

length pseudo-random code unique to each satellite (called the coarse-acquisition code); this is the carrier used by civilian receivers. The L1 signal is also modulated at low frequency to transmit data on the satellite's position, atomic-clock corrections, and status. Both the L1 and L2 signals are modulated with another, much longer (approximately one week per cycle), pseudo-random code unique to each satellite. Encrypted, this code is called the Y-code and is available only to possessors of the decryption key—primarily, the military. Transmitting information on two carriers of different frequency also allows corrections for variations in the speed of electromagnetic radiation through the ionosphere.

There are many potential sources of error in determining the time and location of a user, some already mentioned. Among the largest are how well determined the mathematical inverse problem is (depending on the number of satellites that are visible and the geometry of the satellites relative to the user); uncertainties in the speed of the electromagnetic carrier signals as they pass through the ionosphere (due to ambient electric charge) and the troposphere (due to moisture); the degree to which carrier signals reach the user indirectly via scattering; uncertainties in the broadcast time of each satellite; and uncertainties in the positions of the satellites.

In this monograph, we focus on exploring how variations in the sustainment of the ground segment affect the performance of the system (as measured by user-position error). The ground segment plays a role in monitoring the positions, broadcast time, and health of the GPS constellation. To limit the scope of this pilot study, we focus on the monitoring of satellite position (ephemeris).

Each satellite drifts from its nominal position over time, thus causing deviations between the actual positions of the satellites and the ephemeris data they transmit. Spatial drift from nominal orbits is due largely to gravitational perturbations (e.g., from the Moon and the Sun) and from solar-radiation pressure. The rate of drift varies among the satellite blocks (models), but it is on the order of 1 meter (m) per day.[3] Monitoring these drifts, analyzing the data, and uploading corrections to the satellites are, taken together, one of the functions of the GPS ground segment.

The ground segment has three subsystems: monitoring stations, the Master Control Station, and ground antennas. Monitoring stations passively observe the pseudo-range to each satellite as the satellites pass within their view. Six Air Force (unmanned) monitoring stations are distributed around the globe.[4] These stations are being supplemented by additional stations to be maintained by the National Geospatial-Intelligence Agency (NGA). Together, these monitoring stations will provide a capability to view all the GPS satellites at all times with at least two stations.[5] Data from these monitoring stations are sent to the Master Control Station (MCS) at Schriever Air Force Base, where the data are processed.

Pseudo-range and time data are interpolated and extrapolated with a Kalman filter.[6] The operator compares these data with expected values to determine whether the data received are within limits that are adjusted according to seasonal variation and the expected current satellite-configuration status. If an out-of-tolerance condition or anomaly exists, onboard equipment is first tested for failures to see whether a false reading or transmission of data is the primary cause for the anomaly, rather than the satellite drifting out of the nominal orbital position.

If the satellite is judged to be out of position, the ground operator determines the magnitude of corrections to upload to each satellite, and the priority in which the satellites should receive those updates. The longer that data can be collected by a monitoring station, the more

---

[3] The figure refers to the drift in the estimated range deviation (ERD); roughly half of the satellites (the newer block IIR vehicles) drift less than 1 m per day; the other half drift between 1 and 3 m per day.

[4] Monitoring stations are located at Schriever Air Force Base (Colorado), Hawaii, Ascension (south Atlantic Ocean), Diego Garcia (Indian Ocean), Kwajalein Atoll (Marshall Islands, Pacific Ocean), and Cape Canaveral (Florida).

[5] Interview with Col Kenneth Robinson, SMC/GPG, September 27, 2005.

[6] For an introductory overview of Kalman filtering, see Maybeck (1979), Chapter 1.

monitoring stations that can simultaneously observe a satellite, and the more recent the data, the more accurate the Kalman-filter estimates for the pseudo-range data are. Typically, updates are done approximately daily, but they may be more or less frequent, depending on drift rates.

Data are uploaded to the satellites via the ground antennas, of which there are only four, most collocated with monitoring stations.[7] Figure 2.1 shows the locations of these sites on a Mercator-projection map of the world. These uploads, called *navigation uploads*, are scheduled several days in advance, and the schedules are revised daily. The process of uploading data from a ground antenna takes about 45 minutes (min).[8] As we mentioned above, the corrections to a satellite's position depend on the accuracy of the estimate made of their position from the data the monitoring stations collect. If too much time has elapsed since the collection of those data, the accuracy of the correction becomes suspect and may not reduce error as much as desired. In some cases, if data are not sufficiently fresh, the low confidence level from the Kalman filter might lead MCS personnel to delay an upload until fresh data are received.

With this background in how the GPS functions and is maintained, we now seek to explore how we might model how maintenance practices on the ground segment affect the operational performance of the system.

---

[7] Ground antennas are located at Cape Canaveral (Florida), Ascension Island (south Atlantic Ocean), Diego Garcia (Indian Ocean), and Kwajalein Atoll (Marshall Islands, western Pacific Ocean). There is no ground antenna in Hawaii, but there is an Air Force Satellite Control Network antenna at Pikes Peak, Colorado, that can be used as a backup GPS ground antenna.

[8] Although most uploads are successful, an antenna has occasional problems communicating with a satellite. Sometimes, the problem lies with the satellite, but often the ground-antenna software requires resetting. Resetting ground-antenna software takes about 15 min, delaying the upload. Sometimes multiple resets are needed before an upload can be completed successfully, which, from time to time, prevents a complete upload during the time window when the satellite is within view of the ground antenna.

Figure 2.1
Locations of the Four GPS Ground Antennas

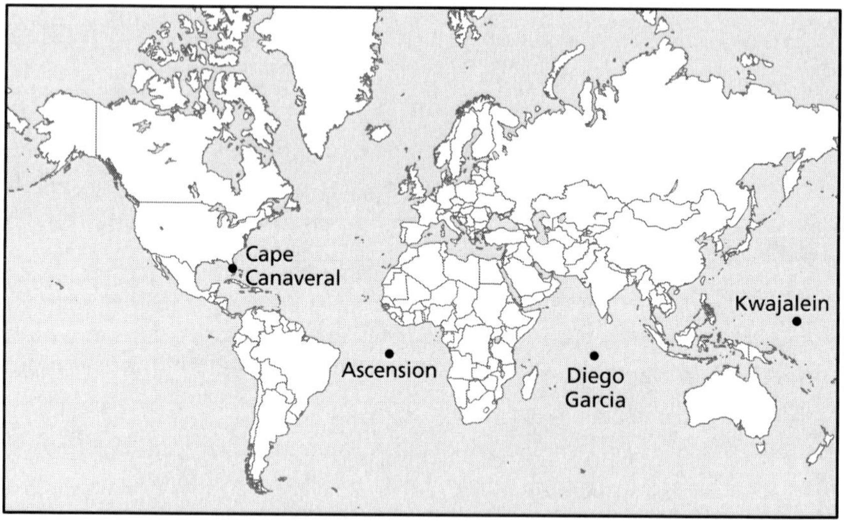

Considerations for Modeling Sustainment Effects on GPS Performance

The starting point for modeling the effect of sustainment activities on operational performance is the selection of a measure of performance. The qualities of the measure of performance determine the scope of the decisions that can be made using the model, and they dictate the minimum level of granularity of the data-collection and analysis efforts needed to feed the model. For the GPS program, regardless of the user, the appropriate sentinel metric of performance is the accuracy of the user's location and time estimates. Here, we focus on location accuracy. Temporal accuracy can be treated similarly. The accuracy goals of the various users will differ, but the type of measure is common to all users. But what exactly do we mean by the *accuracy of the user's location estimate*?

Here we need to distinguish between a model designed to inform programming decisions from one designed to support or to make operational decisions. In both cases, the goal is to model, to the fidelity necessary, a desired objective (performance measure) using the simplest algorithm incorporating the fewest, most economically collected data. But, the objectives of the two model categories differ and, hence, so do the approaches to analysis and type and level of detail analyzed.

Consider first a model developed to make, or to support making, GPS operational decisions. For certain operational decisions, it is significant that the accuracy of a GPS-derived position estimate varies over time and place on Earth (owing to the geometry of the satellite constellation relative to the user). That is, at a given time, a user in New York and one in Sydney may experience differing uncertainties in position accuracy; and a single user may experience position uncertainties that vary over time at the same location. An operator may wish to prioritize corrections to the satellite signals to optimize the performance of the GPS at a given time, or place, or both. Having such an objective requires that the model be detailed enough to analyze these options and that it be supported by comparably detailed data. For example, in this case, idiosyncratic characteristics of individual satellites might prove to be significant and require incorporation into the model.

Now consider the problem addressed in this monograph—how to model the effects of varying sustainment efforts on GPS program performance. In this case, the objective is to inform policy decisions and programming trade-offs. Examples are, What are the consequences to users of the system if sustainment funds are cut by a certain amount? If the life span of a system is to be extended by a number of years beyond its design life, what sustainment support will be needed to meet operational objectives? To achieve the highest returns on capability, how should money be distributed among the options of upgrading technology, increasing the number of ground antennas, and increasing the budget for maintenance? These questions not only are broader but they also address the performance of the GPS program over a longer time period than most operationally focused day-to-day problems.

All these attributes point to a broader measure of performance. In this case, the analyst could use estimates of the temporally and spatially averaged accuracy of a user's position. Or, from the perspective of the space segment, the analyst could use the uncertainties in the ephemeris and time data broadcast by the satellites, averaged over the constellation. We use this latter metric, called the *estimated range deviation* (ERD).[9]

In the absence of corrections, the overall accuracy of the ephemeris and time data of each satellite deteriorates with time. It is one of the principal functions of the ground segment to quantify and correct this drift. How well the ground segment performs this task depends on the reliability of its subsystems: the monitoring stations, the ground antennas, and the Master Control Station. This study explores how to model the effect of the reliability of the ground segment on the accuracy of a user's location estimate as measured by the proxy of the signal-in-space accuracy.

Of these three subsystems, we concentrate on the role of the reliability of the ground antennas in this pilot study, which, because of their limited number, are a good candidate for frequently being the limiting factor in ground-segment capability. The Master Control Station has a backup system, and the number of monitoring stations, especially after being augmented by the NGA stations, gives the monitoring stations more redundancy than the four ground antennas. A full sustainment model would include these other subsystems, but many of the insights from exploring the ground antennas will apply to the rest of the ground segment.

The reliability of the ground-segment subsystems in general, and of the ground antennas in particular, is determined by how frequently the subsystems are unable to perform their mission (the mean time between critical failures) and for how long (mean time to restore function). We group the causes of service interruption into three categories: scheduled maintenance, unscheduled maintenance (breaks), and failures in the communications links between the antennas and the Master Control Station.

Scheduled maintenance includes all maintenance activities that are done on a regular basis, along with installation of system-component

---

[9]   An alternative and related measure is the user range error (URE).

upgrades. These activities have a known schedule and expected dura-tion, but deviations to this schedule can occur because of program slips and for operational reasons. An example of an operational reason might be that, to maintain the overall GPS accuracy in the short term, scheduled maintenance might be deferred to keep as many sub-systems working as possible. This prioritization increases the signal-in-space accuracy over the short term as desired; however, over the long term, such deferments will degrade the system performance because of increased break rates.

Unscheduled maintenance includes hardware breaks, failure of elec-tronic components, and software crashes. Each subsystem is composed of a multitude of parts, each of which will break in time according to some probability density function. Once broken, each component requires some time to be restored to full function. Two factors contrib-ute to this time: the hands-on time to repair the component and the time needed to get the part and maintenance personnel to the location. We call the sum of these times the *time to restore function*, and it also is distributed according to some probability density function.

Finally, the subsystems—in this case, the GPS ground anten-nas—will not be operationally available from time to time because of factors external to the subsystem. One example of such an exter-nal factor is the communications links between the ground antennas and the Master Control Station. Failures in the communications links between this subsystem and the Master Control Station cause a large fraction of this kind of outage. These communications links fall under the purview of, and are maintained by, the Defense Information Sys-tems Agency (DISA); they lie outside of the control of Air Force Space Command. Obviously, changing programming priorities in the Air Force will not improve this situation. Nevertheless, these outages need to be understood quantitatively and be part of any model so that the limits of Air Force actions on the system performance are understood.

To summarize, at a subsystem level, the reliability of the ground antennas has contributions from scheduled maintenance activities, breakage, and communications failures. The service outages from the first contributor are largely under the control of the Air Force, the second are random, and the third are outside the control of the Air

Force. Given that different programming decisions will affect the first two, and the third is beyond the reach of Air Force programming, these three contributions should be treated as separate inputs to the reliability.

Looked at from another point of view, these types of service interruptions reflect the general principle that the performance of a system is not the sum of the performance of its individual parts. Evaluating the performance of a system requires a systemwide view that incorporates not only the performance of the components but also how they mutually interact, how they communicate with one another, redundancies, and the overall command and control of the system. From that view, the evaluation works down from the system to individual maintenance and sustainment activities (a top-down approach).

Bottom-up approaches—such as analyzing the break frequencies and repair times of each component to estimate their probability density functions, and using these functions to analyze reliability of the overall system—play an important role in any complete analysis. But, these approaches alone fail to capture the full system behavior. In the example of the GPS ground antennas, these approaches would not reveal the potential importance of communications-link failures.

Further, the top-down approach naturally focuses the metrics of maintenance and sustainment efforts on the principal objective: how the *users* are affected by maintenance. By starting with the users and working down to the component level, emphasis is placed on the role that a component plays in the overall system and, hence, its contribution to the goals of the user, rather than focusing on how well the component performs its own *distinct* function. In this manner, the top-down approach successfully links maintenance and sustainment activities with the operational function of the system.

These merits of a top-down approach do not invalidate the need for an understanding of component-level failures. Rather, the system-wide, operational view places the components in context and reveals a priority of data collection and analysis. That is, a system view indicates which subsystems or components are most problematic and therefore deserve the highest level of attention in failure-and-repair data collection and analysis. Once the key problems are identified, whether they are component failures, communications-link failures, lack of redundancy, or other issues, data for costs to remediate the problems can be estimated by examining these service-interruptions modes in detail. This detail ties dollars invested to overall system performance as measured by the users' needs.

A complete, predictive analysis of maintenance and sustainment efforts for space systems then unfolds in the following steps. First, the operational objectives of the users are quantified in a way that reflects the long-term behavior of the system likely to be affected by programmatic decisions. These *operational* objectives then define the metrics for *maintenance and sustainment*. A predictive model based on a systemwide view links the maintenance and sustainment efforts to the operational metrics. This predictive model then reveals critical problem areas, which can be explored in greater detail. Once the critical areas are identified, additional analysis at the component level then links these remedies with costs, indicating how investments in resources affect operational performance.

For these reasons, in this monograph we start with a top-down approach to modeling the GPS. Although the scope of this study constrained us from linking this work to component-level analysis and, hence, directly to costs, the approach explored in this monograph complements ongoing component-level analysis being done by AFSPC/A4S. Linking the analysis presented in this monograph with ongoing work at AFSPC can present a complete, predictive model of space systems that reveals how dollars allocated in the budget affect the operation of the overall space system in terms of operational performance.

Evaluating system performance from a top-down perspective should not be confused with a metric that captures how often the system performs nominally. For most space systems, redundancy keeps them operating nearly always. We need to know when the system will break, and what will make it break, without doing the experiment. Hence, a model is needed that captures the system-level behavior. For this pilot study, we built a model of the GPS ground-antenna subsystem that predicts when the system will fail.

For this reason, we explore how the variation of the frequency of service interruptions and the time to restore function affect overall

system performance. We do not attempt to evaluate the current operation of the system. Indeed, the complete data set needed to do that
analysis may not be available. First, we were not able to locate data for
how often and for how long ground antennas are not mission-capable
because communications links supplied by DISA have failed. Although
DISA may collect these data, they do not appear to be collected or
reported in the Air Force. Further, we were unable to locate data for
how often software systems crash, thus impeding or preventing a navigation upload. Anecdotal information indicates that this problem may
be sufficiently frequent to warrant collection and analysis.

Henceforth, we explore how variations in these failure rates and
durations in the ground antennas will impact GPS program performance. To capture the stochastic nature of the frequency of antenna
failures, whatever their cause, we use the exponential probability distribution function for mean time between failures, such that the probability p as a function of time t is

$$p(t) = \theta \exp(-\theta t) \tag{2.1}$$

where the mean time to failure is θ^{-1}. An exponential distribution arises
from an expectation that failures occur randomly according to a Poisson process. Hence, it does not capture phenomena for which failure
rates vary with time, which can happen when failures increase with
age of the components, or in the case of software crashes, occur more
frequently earlier in their implementation. We also use the exponential distribution to model the probability density function of the mean
time to restore function.

Other distributions could equally well be employed for these
random variables,[10] including the two-parameter gamma distribution[11]
and the three-parameter Weibull distribution.[12] These distributions
can express the nature of the variance of the distributions more flex-
ibly than can the exponential distribution, and they can capture failure

---

[10] See Mann, Schafer, and Singpurwalla (1974), Chapter 4, for a discussion.

[11] Mann, Schafer, and Singpurwalla (1974), pp. 259–264.

[12] Weibull (1951); Mann, Schafer, and Singpurwalla (1974), pp. 184–258.

rates that vary with, for example, the age of the component. But, given the limited data available to fit these distributions and that we expect the mean times of both failures and restoration to dominate the analysis over their variances, we have opted to use the simpler exponential distribution.

CHAPTER THREE

A Predictive Model for the Sustainment of GPS Ground Antennas

In this chapter, we outline a pilot model for the GPS ground antennas, developed in accordance with the themes outlined in Chapter Two. The model is then applied to show how it can be used to predict some of the circumstances under which the GPS system will fail its users as a result of service interruptions of the ground antennas.

A Pilot Model

Two attributes of the problem strongly suggest a stochastic simulation rather than a deterministic calculation. First, the objective itself, the error in the signal in space, is a random variable with a probability density function over time, and the variance of this distribution is of interest. Second, the input data—the reliability of the ground-segment subsystems in terms of mean times between failures and mean times to restore function—are also random variables. Further, under current practices, corrections to the signal in space are scheduled as discrete satellite uploads—typically daily, rather than continuously. This aspect of the problem indicates a discrete rather than a continuous model. The approach pursued here, therefore, is a discrete stochastic simulation of the GPS over time.

The principal goal of the simulation is to explore how changes in the availability of ground antennas affect the performance of the GPS program. In this monograph, we use the simulations to examine changes in maintenance performance and the addition of a fifth antenna. The algorithm's flexibility allows a broader range of issues to be analyzed, including launch delays of new satellites, loss of antennas, and changing technologies that reduce drift in the system.

The simulation algorithm is hosted in Microsoft Excel and written in Visual Basic for Applications. The Excel shell holds input data, provides the user interface, and presents the results of the calculations. Supporting data are held in several supporting files, and calculation results are stored in output files in text format. Figure 3.1 depicts the algorithm's execution flow.

A model run begins by generating a time sequence of when the ground antennas are functioning and when they are not (downtime). This sequence is generated stochastically for each ground antenna by causing service outages according to a Poisson process for the mean time between critical failures (MTBCF) and for a duration of outage (mean time to restore function, MTTRF) according to an exponential distribution. The model uses the random-number generator in Excel's Visual Basic for Applications for the stochastic contribution. Once generated, this sequence of downtimes is fixed for a given simulation run.

Next, the algorithm loads a schedule of when each satellite is visible to each ground antenna. This file was calculated in advance of the simulation and is stored in an input file. These two steps lay the groundwork for the actual simulation. The simulation begins at zero time and advances a model clock by 45 min each step (i.e., 32 steps per day). At each time step, the ERD deteriorates by a fixed amount unique to each individual satellite. Data for this drift rate are from AFSPC–provided historical observations in the form of daily drifts in the error over a week. These errors grow approximately linearly over time, with some scatter; we fitted a linear curve to interpolate the ERD data for each satellite.

At the end of each model day, the algorithm creates a prioritized schedule for uploading ephemeris and time data to each satellite. First, the satellites are sorted from greatest error to least error. Starting with the satellite with the greatest error and proceeding down the sorted list, the algorithm searches for a ground antenna in the soonest time step

Figure 3.1
Flow Diagram of the Simulation Algorithm

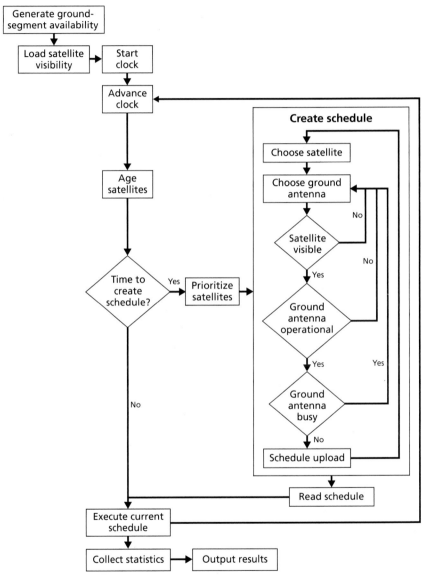

that is operational (not in a downtime), available (not busy uploading to another satellite), and visible (to that satellite). It advances through the ground antennas for the first time step until it finds one that meets all three requirements. If it finds none, it goes to the next time step, doing the same, until it exhausts all time steps in that day. If it cannot schedule an upload for that satellite on that day, the upload is deferred until the next day's scheduling, and the satellite's ephemeris and time errors remain uncorrected. The model proceeds through all satellites in this manner.

After a schedule is created, the model advances the model clock and increments the satellite ERD. For each time step, if an upload is scheduled, it resets that satellite's ERD to zero. During this simulation, the algorithm collects and outputs to text files relevant computations, including but not limited to satellite ERD, average ERD over the entire satellite constellation, ground-antenna state at each time step, and ground-antenna usage (uploads per day). Selected data are also presented graphically. All statistical analyses of the ERD discard the first model month simulated, thereby ensuring sampling of steady-state behavior (i.e., to avoid transient, or spin-up effects of the simulation).

Illustrative Calculations

This simulation captures the desired features described in Chapter Two. Here, we examine a sampling of the programming decisions that such a model can address. Because the data mentioned above are fragmentary, we present these results to illustrate trends and applications, not to address current programming decisions. Hence, we do not run the simulations for single values of the MTBCF and the MTTRF. Rather, we explore how variations in the values of these random variables affect system performance. In the next chapter, we discuss other possible applications of the model, what steps need to be taken to implement a tool such as this one, and some generalizations about those aspects of this analysis of the GPS program that extend to other space systems.

We describe two sets of analyses: The first explores how the performance of the GPS program deteriorates as the MTBCF and

MTTRF worsen; the second explores programming trade-offs between improving either the MTBCF or the MTTRF and adding (or losing) a ground antenna to the current set. In both sets of calculations, we let the MTBCF and MTTRF vary as parameters. Also, for both sets of results, we use the average ERD across the satellite constellation as a measure of system performance, which, to be concise, we call ψ. The value of ψ varies with time. For the purposes of this analysis, it is a function of the reliability of the ground segment. As mentioned above, ψ is a random variable in time.

Current Antenna Configuration

Figure 3.2 shows the probability density function of ψ for three simulation runs for three different MTTRF at a fixed MTBCF (50 hr). The first 1,000 data were rejected to avoid spin-up bias; over 10,000 data make up the histograms. In the simulations depicted in the figure, maintenance performance deteriorates by an order of magnitude: from 5 hr on average to restore the subsystem to full function in the lower panel to 50 hr in the upper panel.

Note, first, that despite the large sample size, all of the distributions deviate from a normal distribution—most notably, in having positive skewedness (steep left side and significant tail on the right side), which increases as the MTTRF increases. This asymmetric spreading of the distribution is most noticeable in the range of the distributions, which increases by nearly a factor of 5 as the MTTRF increases by a factor of 10. It is a reflection of the robustness of the GPS that the means of the distributions change only by a factor of 1.2 over this range of MTTRF. From a user's viewpoint, this robustness denotes that, as the constellation's ephemeris and time data worsen, the increase in the fraction of users experiencing a location (or time) error greater than a given threshold will change more than the mean error (over time) experienced by those users.

To evaluate how operational performance changes with quality of maintenance performance, we seek a sensitive measure for a sentinel metric, one that reflects both the spreading of the distribution and the small, but important, shift in the mean: To capture these attributes, we

Figure 3.2
Histogram of Average ERD for Three Different MTTRF

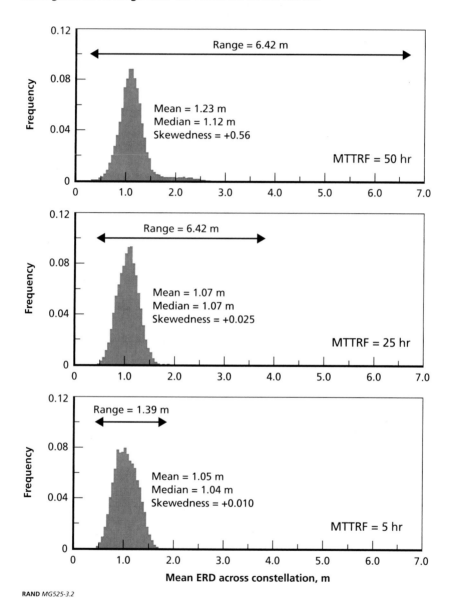

use the 99th percentile of the distribution (i.e., the level of error for which 99 percent of the observations fall below). The percentile will shift as the mean shifts, capturing this effect, while also being less susceptible to outliers than the range.

We now examine how the MTBCF, the MTTRF, and the number of antennas affect the 99th percentile of ψ. Before showing these results, we remind the reader that the relative value of a ground antenna depends on what fraction of a satellite's orbit is visible to that antenna. This fraction is a function of how close to the horizon an antenna can communicate with the satellite, and the altitude and inclination of the orbits. Over time, the fraction of an orbit visible to an antenna will be dependent on latitude and independent of longitude (unless the orbital and rotational periods are exactly in phase), as shown in Figure 3.3.

Figure 3.3
Visibility of GPS Satellites by Ground Antennas as a Function of Latitude

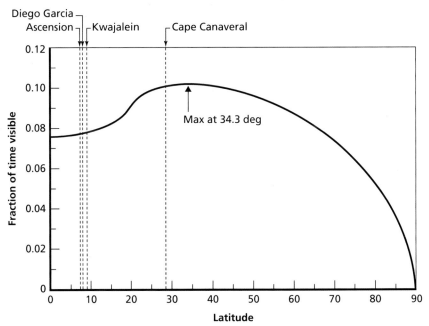

Figure 3.3 is for GPS ground antennas, assuming (for argument) that they can effectively communicate at all angles down to 5 degrees (deg) from the horizon. The maximum visibility, and, hence, all other factors being the same, the optimal position, lies at latitude 34.3 deg north or south, equally spaced in longitude around the Earth. Hence, a given MTBCF or MTTRF for one antenna will have a different effect on the overall system than that same MTBCF or MTTRF for another antenna. The simulation captures these positional dependencies.

Figure 3.4 shows how the 99th percentile of ψ varies with maintenance performance on the four existing ground antennas. The lower panel shows varying MTBCF at fixed MTTRF (5 hr); the upper panel shows varying MTTRF at fixed MTBCF (50 hr). The values selected for the MTBCF and MTTRF were chosen *at random* within the domain of interest to diminish the likelihood of resonance.[1] Separate stochastic runs were performed for each antenna, but each datum represents a common mean of the exponential distribution used to generate the time sequence of MTBCF or MTTRF for each ground antenna.

Looking first at the lower panel of Figure 3.4, we can see that, for MTTRF = 5 hr, improving maintenance performance so that the MTBCF exceeds about 15 hr provides diminishing returns to the user. The plot also indicates that looking only at the overall GPS performance does not provide insight into lurking problems caused by the MTBCF; overall system performance is an anemic measure of latent maintenance problems for MTBCF >15 hr. That is, without modeling similar to that shown in Figure 3.4, maintenance performance could deteriorate leftward on the figure without forewarning of the impending drop in performance for MTBCF <10 hr.

---

[1] The GPS has natural periodicity. The orbits are nearly (but not exactly) 12 hr, the navigation uploads are nearly daily, and the time for an orbit to precess around the Earth is of a considerably longer period. If the MTTRF and MTBCF were spaced evenly in time at a frequency that was a resonance frequency of the GPS, it could bias the results. Selecting the sample spacing randomly suppresses this effect.

Figure 3.4
Effect of MTTRF and MTBCF on ERD for Current Ground Antennas

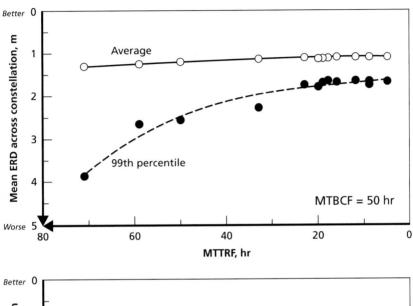

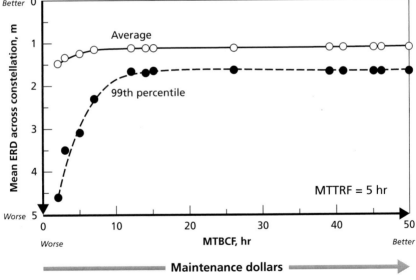

It is no mystery why this is so: Subsystem redundancy masks underlying maintenance problems that are revealed only when redundancy fails—in this case, at about MTBCF = 15 hr. Here, the choice of a monitoring metric is important, but predictive modeling provides the clearest insights into the future health of the system.

Looking now at the upper panel of Figure 3.4, we see that the MTTRF is a much more sensitive, or sentinel, indicator. That is, variations in the MTTRF affect the overall GPS performance in a more continuous manner than the MTBCF curve does. Taking more than about a day to restore the system function from breaks that occur, on average, every 50 hr significantly increases the scatter in the ERD.

Figure 3.4 shows how variations in MTBCF (MTTRF) at fixed MTTRF (MTBCF) affect the 99th percentile of ψ. The contour plot in Figure 3.5 shows how the 99th percentile of ψ changes as MTBCF and MTTRF co-vary. The contour plot shows that neither the MTBCF nor the MTTRF alone is a suitable indicator of maintenance performance, nor is either alone a good forecaster of future performance. A particular value of MTBCF (MTTRF), for example, can also be a disastrous value, depending on the simultaneous value of the MTTRF (MTBCF).

These observations emphasize the point that it is the operational objective—in this case, the 99th percentile of ψ—that serves as the best measure of the health of the overall maintenance performance. A predictive model, such as that described in this monograph and used to generate Figure 3.5, is needed to shed light on what a pair of MTBCF and MTTRF values means to the operator or warfighter. That is, whether a pair of MTBCF and MTTRF values forewarn that the system is in peril or assures that satisfactory operation is secure for at least the near term.

Figure 3.5 also introduces another important point. Similar operational system performance can be achieved by more than one pair of values of MTBCF and MTBCF (i.e., by any of the values along one of the contour lines in Figure 3.5). After deciding on the desired contour, or bounds of contours, that the maintenance needs to uphold,

Figure 3.5
Contour Plot of the 99th Percentile of ψ as a Function of the MTTRF and the MTBCF

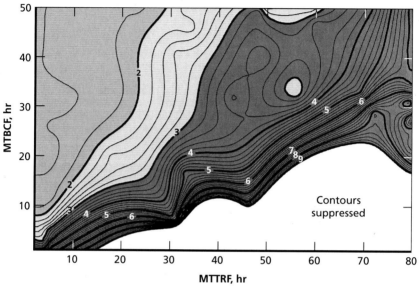

the analyst can see that the most desirable position along that contour depends on the cost trade-offs between achieving a low MTTRF or a high MTBCF. The ultimate expression of quantifying these trade-offs would be to overlay contour lines of cost on Figure 3.5. Then, optimal values for MTTRF and MTBCF might be determined by noting where the minimum cost falls along the contour of desired operational performance.

A full treatment of how to negotiate these cost trade-offs is beyond the scope of this monograph. The economics of public goods does provide useful guidance and context for cost trade-offs among maintenance investments and other programmatic areas for such services as GPS and other space capabilities (see the Appendix).

Note that the above analysis is of the steady-state behavior of the GPS. That is, we explored how the GPS system would perform over the long term if the MTBCF and MTTRF of the ground antennas took on a specific value (within a wide range). We did not examine the *relaxation time* of the system to varying perturbations—how fast the system would transition from one performance state to another upon a change in either MTBCF or MTTRF of this subsystem. For example: If a ground antenna's maintenance performance declines abruptly (or is lost completely), how long does the system take to respond to this shock?

Understanding the relaxation time of the system is just as critical as understanding how the system will perform in the steady-state under certain maintenance practices. The algorithm described here can perform this perturbation analysis, but this work could not be accomplished within the time frame of this study. Using plots similar to Figure 3.5 to make programming decisions would benefit from the knowledge of the relaxation behavior of the system. If the pairs of values of MTBCF and MTTRF begin to migrate up contour lines of operational performance, how long does a maintainer have to rectify the situation before performance drops below an acceptable level?

Alternative Antenna Configurations

The analysis can be used to examine programming trade-offs as well. One (hypothetical) example is whether to invest in maintenance activities or to add another antenna. We arbitrarily added to the model simulations a ground antenna at the monitoring station site in Hawaii (latitude 21 deg north) to see how this addition affected the system performance relative to maintenance activities. Figure 3.6 shows the results of these calculations for the 99th percentile of ψ. The addition of an antenna does not have a statistically meaningful effect with the variation in MTTRF, and it may permit a slightly lower threshold at which MTBCF is critical.

Now consider the loss of an antenna from the system, perhaps as a result of a natural disaster or deliberate sabotage. Loss of an antenna has different characteristics from the addition of an antenna. In a satisfactorily functioning system such as the GPS, adding an antenna adds redundancy more than it adds capability, and additional redundancy does not add much to performance in a well-maintained system (Figure 3.4). But losing an antenna can potentially wreck the capability and,

Figure 3.6
Effect of MTTRF and MTBCF on ERD for Five Ground Antennas

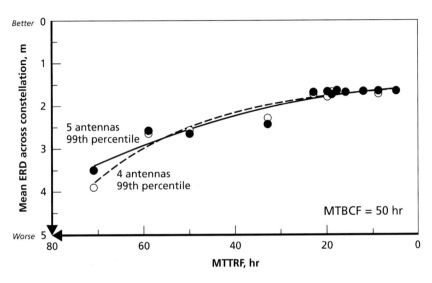

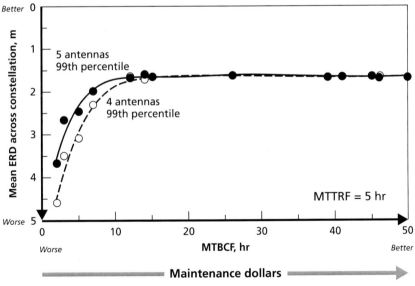

hence, the performance. Figure 3.7 presents simulation runs for three ground antennas (in this case, loss of the antenna at Ascension Island). Clearly, having only three antennas impacts performance for all but the best maintenance performance. Here, if MTBCF is less than about 20 hr (and restoration is achieved, on average, in 5 hr), the GPS performance deteriorates. And, if the MTTRF is longer than about 10 hr (and the MTBCF is 50 hr), the performance deteriorates. Hence, to sustain system performance with only three antennas requires exemplary maintenance on those antennas. Again, Figure 3.7 indicates the steady-state performance of the system with three antennas and does not disclose how long it would take to transition from the four-antenna steady state to the three-antenna steady state.

Figure 3.7
Effect of MTTRF and MTBCF on ERD for Three Ground Antennas

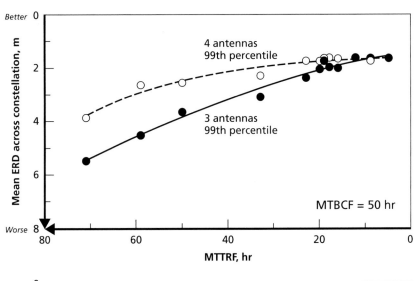

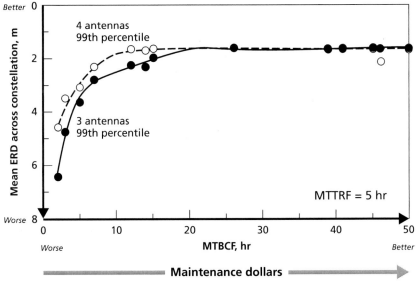

CHAPTER FOUR

Conclusions and Next Steps

In this monograph, we have presented some of the power of a model that links GPS operational performance with measures of unscheduled maintenance activities (of both the antennas themselves and their communication links to the Master Control Station) and have discussed some of the basic features that such a model should possess. The spirit of the model can be expanded to embrace all contributions to service interruptions of the ground segment. In particular, we argued for modeling with a systemwide view, rather than a bottom-up approach that builds on data of failures and repairs of individual components. The systemwide view places these components in a context, and the component-level data provide the link between the operational metric and the dollars invested in maintenance and sustainment. Further, we distinguished the differences of this type of model from one with operational objectives, emphasizing that the choice of performance metric is central to the analysis. What, then, are the next steps toward implementing such a model?

First, the model would need to be expanded to include the reliability of the monitoring stations, the Master Control Station (and its backup facility), and the communications network that links all these subsystems. The GPS program is a system, and like many complex systems, its performance is not that of the sum of its parts. The components interact. It is possible to have a system composed of well-functioning parts, but not have a well-functioning system if the parts are not well integrated. Although counterintuitive, it is also possible to have a well-functioning system composed of poorly functioning parts, if the system has adequate redundancy and excellent command and control. For these reasons, we selected a metric of performance that reflects the overall *system* performance, and not one that focuses on the performance of the ground antennas. But, to fully understand the role of maintenance activities on the system, all the major components (monitoring stations, Master Control Station, and communications links) and how they interact need to be added to the analysis. That is, the work must grow from a subsystem analysis to a system analysis.

Beyond system effects, if other major components, such as the monitoring stations, are included, two added levels of complexity are introduced. The model must be able to estimate the quality of the Kalman-filter estimate of the error in the signal in space as functions of how long a monitoring station observes a satellite, how many monitoring stations can do so simultaneously, and the age of the data. With the addition of NGA monitoring stations, the model must include this added capability, but it must also assess any risks of relying on resources outside of the Air Force.

Second, there is a continuing need for comprehensive data on when each of the subsystems is not functioning well enough to perform its assigned mission. These data do not appear to be reported within AFSPC in a form necessary to support the analysis described in this study and cannot in general be deduced from data on component-level failures. This collection effort should include instances when the software crashes and needs resetting, as well as such factors as failures of the communications links, even if these lie outside the control of AFSPC, and any other times (of which we are unaware) that a subsystem is operationally unavailable.

The trigger for when a service interruption, such as a software failure, gets reported should reflect its effect on the system's operation. Recurring, short software interruptions that require software resets can preclude a navigation upload if they are frequent enough to prevent a ground antenna from operating long enough to perform the upload while the satellite passes within view, even though each one of these interruptions might be short enough that none triggers the current collection system for reporting outages. Even if external factors play a dominant role in the ground antennas' reliability, this dependence is important to know quantitatively.

Third, once analysis at the subsystem level, based on sentinel metrics expressed in operationally relevant terms, reveals the problematic areas of the system—which sets of components are most fruitful to understand in detail, thereby economizing on time and resources—data collection and analysis can be extended to the component level in a targeted fashion. These targeted components can be studied for their individual break rates, repair times and costs, and replacement costs. Using this approach, the operational metrics can be linked to programming decisions, such as buying more spare parts, reducing lead times for repair, or training more maintainers.

Resource and time limitations have prevented us from exploring these linkages, but they need to be estimated for the various classes of failures before a model like that described here can be used directly by programmers to link dollars to readiness. Key issues are, What causes breakages of mechanical components, failures of electrical components, and software crashes? Specifically, are system failures correlated with service cycles, duration of use, or other factors? And, what are the consequences of deferring scheduled maintenance on these systems to future break rates and break types?

For many components managed by AFSPC, these issues appear to be poorly understood, although they are reasonably well understood in the realm of aircraft, where most engine maintenance correlates with flying hours and brake and tire maintenance correlates with take-offs and landings. Unfortunately, space system components and subsystems present challenges beyond those encountered in the analysis of aircraft. Space systems have fewer identical components, and fewer systems in general, on which to collect statistically meaningful data. Also, past performance of how software modifications affect the mean time between failures and how long those failures take to repair is not clearly a good approximation of future behavior. These important issues require a separate study.

Of course, by looking at the reliability of ground antennas alone, we have restricted the analysis. Performance can be improved by other types of efforts, such as improving the quality of the GPS's algorithms, introduction of more-advanced technologies, and cross-link capability among the satellites. Cross-link capability would allow a single ground antenna to communicate navigation uploads for all the satellites to a single satellite, which would, in turn, pass the navigation uploads on to the other satellites. It would decrease the number of ground antennas needed, thus decreasing the vulnerability of the antennas to sabotage. Yet, introducing new technologies to improve system performance may introduce new reliability issues, especially in the software area, potentially decreasing system performance. Hence, program decisions are best made with a wider, quantitative view of the consequences of trade-offs among sustainment, additional hardware, and technology advances. A model such as the one proposed here could include these trade-offs.

Fourth, the analysis should be expanded beyond the steady state to examine the relaxation times of the GPS due to maintenance-performance perturbations. All the simulations in this monograph were performed at steady state. How long the system takes to respond to changes in ground-segment maintenance performance are vital for understanding the lead times needed to rectify latent problems before they become real problems. It also aids in understanding historical data on perfomance metrics. If values of the maintenance-performance measures of the ground segment, such as the mean times between failure and mean times to restore function, are on the order of, or longer than, the relaxation time of the GPS overall, then the system performance will naturally oscillate. Studies of relaxation times of the GPS would help interpret such oscillatory behavior, or, if desired, avoid it.

Finally, we make a few observations on what lessons can be gleaned for the analysis of other space systems. Many other space systems have characteristics that make them more complex to analyze. For example, such systems as range support have more users and many more unique, or nearly unique, components than the GPS system. These qualities further complicate the selection of a performance metric, and they pose additional obstacles to collecting meaningful data on system reliability. Nevertheless, the general principles discussed in this monograph are applicable—notably, how to select a sentinel performance metric, the appropriate level of analysis, and the priorities for data collection on subsystem reliability. We hope that this overall framework will contribute to a better predictive understanding of how sustainment investments affect space system readiness.

The GPS as a Public Good

The GPS signal that results in geographical measurements is of use to many different users for differing purposes. The GPS signal is a classic case of a public good. One user's consumption of the signal does not diminish another user's consumption. There can be ten, a hundred, a thousand, or millions of users. Each user is consuming the signal without appropriating it for exclusive use. This phenomenon is important to resource-allocation decisions, because classic reliance on free-market economics results in a market failure, which we will explain below.

Consider two firms, User A and User B. Their demands for the output of GPS are depicted in Figure A.1. Note that each demand curve has the normal downward-sloping shape. As the price falls, more services are demanded. These curves are derived from the cost-minimizing (profit-maximizing) behavior of firms. However, in a defense context, such demand functions are best thought of as minimizing the costs of producing a given level of capability. At high prices for GPS services, relatively little is demanded because other factors of production are used as substitutes. The market failure comes into being because the horizontal summation of these demand curves does not take into account the public-good nature of the capability. If User A pays for a certain amount, User B can use the good for free, or vice versa. Hence, each user tries to "free-ride" on the other's purchase and the market produces too little of the good.

Instead, the two demand curves should be summed vertically, as depicted in the lowest panel of Figure A.1: To make the market

Figure A.1
Schematic of Demand Curves for a Public Good

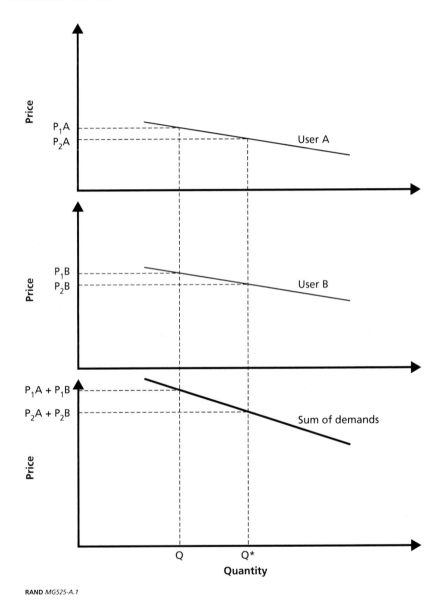

work, the price each user is willing to pay at a given quantity should be
added and then compared to the marginal cost of the service. At equi-

librium, the sum of the prices should equal the marginal cost of providing the services. Figure A.2 shows a GPS supply function (defined by the 99th percentile of the average ERD) overlaid on these demand curves and the resulting equilibrium.

We have used the term *prices* here to reflect what users would be willing to pay as a function of their cost-minimizing calculations, which take into account the costs of substitute factors of production. It is easier said than done to actually construct these demand functions and use them in resource-allocation decisions.

However, and interestingly, cost-sharing relations, as long as the weights sum up to one, can have the same effect. The end result of each user's calculations will be to equate its share times the cost charged at various levels, which will ensure that the sum of the prices will equal the marginal costs at the level of output produced. This result is very similar to the reasoning behind NATO infrastructure burdensharing,

Figure A.2
Demand Curves and Supply Curve for GPS Public Good

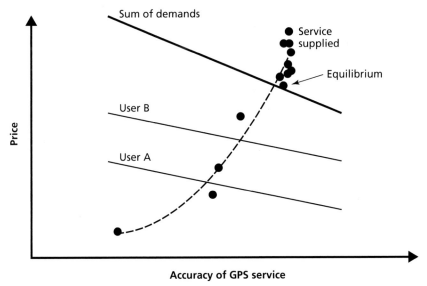

an operational and historical example of an application of this sort of "public goods" reasoning.

We are not recommending that complete demand curves, such as those discussed above, be calculated; however, we are suggesting making some estimates of aggregate demands and applying the reasoning to GPS services. Program Objective Memorandum (POM) and budget deliberations would be the proper vehicles for such estimates. The GPS is used in many weapons, weapon systems, and other applications. For example, such programs as the Joint Direct Attack Munition (JDAM) do not contribute to the GPS program. Yet, without GPS, JDAM would have no utility. Enumerating such programs and calculating what contributions would make sense to ensure appropriate levels of GPS services is clearly an exercise that would make sense in making POM decisions. In addition, since non–Air Force programs would be included in the calculations, decisions made at the level of the Office of the Secretary of Defense should be better informed.

Bibliography

Mann, Nancy R., Ray E. Schafer, and Nozer D. Singpurwalla, *Methods for Statistical Analysis of Reliability and Life Data*, New York: John Wiley and Sons, 1974.

Maybeck, Peter S., *Stochastic Models, Estimation, and Control*, Vol. 1, New York: Academic Press, 1979.

Weibull, W., "A Statistical Distribution Function of Wide Applicability," *Journal of Applied Mechanics*, Vol. 18, 1951, pp 293–297.